T·H·E B·E·S·T O·F
APPLES

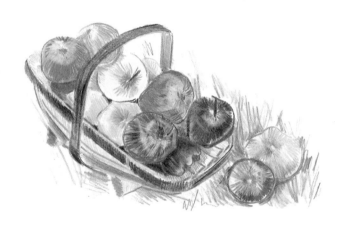

T·H·E B·E·S·T O·F
APPLES

GARAMOND

First published by Garamond Ltd, Publishers
Strode House, 44–50 Osnaburgh Street, London NW1 3ND

ISBN 1–85583–051–5

Illustrations by Madeleine David

Printed and bound in UK by MacLehose and Partners Ltd.
Typeset by Bookworm Ltd.

CONTENTS

INTRODUCTION

Of all the fruits that grow in the temperate world, the apple is the undoubted king. The moist, cool climate seems to bring out all of its sharp flavour in a crisp, juicy texture, and the farther south one goes, the softer and mealier the apples become and the blander their sweetness.

The apple is thought to have originated in the mountains of the Caucasus and its cultivation goes back to the Stone Age, even farther if the story of Adam, Eve and the Tree of Knowledge is to be taken literally. Some think that it was not an apple tree at all that was involved but another fruit, possibly an apricot or a fig, but the majority still come down in favour of the apple.

The Ancient Egyptians cultivated apple trees and the Greeks and Romans knew them well. By the end of the fourth century AD, there were thirty-seven varieties of apple on record. From then on, wherever western civilization went, apples went too – to the New World, to the Antipodes as well as throughout the Old World, to settle wherever the climate and growing conditions suited it. Today, there are well over a thousand varieties of apple in existence but the search for perfection continues unabated. Commercial growers have still to find the perfect apple, one that is a heavy cropper and resistant to disease, of equally good flavour and appearance, unaffected by packing and lengthy transportation.

The apple's versatility cannot be overemphasized. It can be enjoyed at any time of day, in any course of a meal. A fine French cuisine, that of Normandy, relies heavily on apples – fresh, cooked with all kinds of meat, poultry and game, or sliced into large open tarts. The French also have a magnificent apple brandy called Calvados. Americans have apple jack and the British are famous for their cider. Last but not least, there is cider vinegar, which is infinitely preferable to some harsh, inferior potions that go by the name of wine vinegar.

As far as the cook is concerned, apples fall into two main categories, dessert or eating apples, and those suitable for cooking. However, there is no reason why you should not explore the flavours and textures of dessert apples for sauces, pies and other apple dishes as well.

When choosing apples in the store or on a stall, go for fresh, firm fruit with a smooth, shiny skin and good colour. A pale, anaemic colour is usually a sign of immaturity and consequent lack of flavour, which is unlikely to improve, no matter how long you wait. A slightly shrivelled skin could be caused by incorrect or overlong storage. An over-ripe apple, which also tends to be over-size, will be mealy and rather soft to finger pressure, and its flavour will be impaired.

Apples are best frozen as pie slices or purées or apple sauces. Blanch slices in acidulated water and dry or sugar freeze. They can also be dried.

Apples are not just for eating by themselves, for cooking with, or for making wine. They are also very good for you and 'an apple a day keeps the doctor away' is an old saw which is based on scientific fact. Apples, particularly the peel, are rich in vitamins and contain at least six essential minerals, as well as malic acid, which is a powerful aid to digestion. A raw apple is good for the teeth and gums and sweetens the breath; eaten last thing at night it will help you to sleep well. Apple pips can be poisonous, but only in considerable quantities – a cupful has been known to be lethal.

POLISH HERRING
APPETIZER

450 grams (1 lb) salted herrings
2 tablespoons, finely chopped mild onion or shallots
6–8 tablespoons coarsely chopped, peeled apple
8 tablespoons thick sour cream
2 tablespoons lemon juice
½ tablespoon sugar
1–2 tablespoons finely chopped parsley, chives or green spring onion tops, to garnish

Desalt the herrings for 48 hours as follows. Place them in a large bowl in the kitchen sink and let cold water trickle over them constantly until required; or, alternatively, cover the herrings with cold water and leave for 48 hours, changing water frequently – several times a day.

Drain the herrings thoroughly. Fillet them, discarding all skin and bones, and cut each fillet into bite-sized pieces. Arrange the pieces in a shallow serving dish.

In a small bowl, mix the onion or shallots and apple with the sour cream, and flavour to taste with lemon juice and sugar. Pour over the pieces of herring and mix lightly but thoroughly to coat them completely. Scatter the top with parsley, chives or spring onion tops and put aside in a cool place for a couple of hours to develop flavours before serving.

Serves 6

HOT APPLE SOUP

700 grams (1 ½ lbs) apples
One 5 centimetre (2 in) cinnamon stick
Twist of thinly peeled orange rind
25 grams (1 oz) potato flour or cornflour
About 300 millilitres (½ pint) orange juice
125–175 grams (4–6 oz) sugar
Lemon juice
300 millilitres (½ pint) hot single cream
Ground cinnamon or cinnamon croûtons, to serve

Wash, quarter and core 450 grams (1 lb) apples. Put them in a pan with the cinnamon stick, orange rind and 600 millilitres (1 pint) water. Bring to the boil and simmer, uncovered, until the apples have disintegrated. Rub the mixture through a sieve, discarding the cinnamon stick and rind. Pour the soup back into the rinsed-out pan.

Blend the potato flour or cornflour smoothly with a few tablespoonfuls of orange juice. Reheat the soup and when it is halfway to the boil, stir in the potato or cornflour paste, together with the remaining orange juice. Sweeten to taste with sugar, bring to the boil, stirring, and simmer gently for 3 or 4 minutes.

Peel and core the remaining apples. Grate them into long shreds on the coarsest side of a grater. If the soup is not to be served immediately, squeeze a little lemon juice over them and toss lightly with a fork.

Just before serving, bring the soup to simmering point. Stir in the grated apples and cream. Taste for sweetness and serve, each portion garnished with either a pinch of cinnamon or a few cinnamon croûtons.

Serves 4

BAKED CHICKEN
WITH APPLES

1 roasting chicken, about 1.6 kilograms (3½ lbs)
50 grams (2 oz) butter
3 slices lean unsmoked bacon, diced
700 grams (1½ lbs) firm, aromatic dessert apples
Salt and freshly ground black pepper
300 millilitres (½ pt) dry cider

Clean the chicken carefully and cut it into 8 or 9 neat pieces. Dry the joints with paper towels and put them aside until needed.

In a large, heavy frying pan, melt the butter until foaming but not coloured. Add the diced bacon and fry gently until limp and golden but not crisp. With a slotted spoon, drain bacon bits and put aside on a plate until needed.

In the fat remaining in the pan, sauté the chicken joints over moderate heat until a rich golden colour on all sides.

Quarter, core and slice the apples vertically. (I prefer not to peel them as the skin helps the slices to keep their shape.)

In a heavy, ovenproof casserole with a tight-fitting lid, arrange a third of the apple slices. Lay half of the chicken pieces on top and scatter with half of the bacon pieces. Season lightly with salt and freshly ground black pepper. Repeat with another layer of apples, the remaining chicken joints and bacon pieces, season lightly and top neatly with the remaining apple slices.

Pour the cider into the juices remaining in the frying pan, and, over a moderate heat, scrape the bottom and sides of the pan clean with a wooden spoon. Pour the simmering liquid all over the top layer of apples. Cover the casserole tightly and bake in a moderate oven (375°F, 190°C, Mark 5) for 40–45 minutes, or until the chicken is tender. Serve straight from the casserole.

Serves 6

PERSIAN STUFFED APPLES

8–10 large, tart apples
Butter
Ground cinnamon (optional)

Stuffing
5–6 tablespoons yellow split peas
1 Spanish onion, finely chopped
25 grams (1 oz) butter or oil
450 grams (1 lb) lean minced beef
½ teaspoon ground cinnamon
Salt and freshly ground black pepper

Basting sauce
4–6 tablespoons cider vinegar
2–3 tablespoons sugar

Start by preparing the stuffing. Boil the split peas in unsalted water until they are very soft, about 30 minutes. Meanwhile sauté the onion in butter or oil until soft and lightly coloured. With a slotted spoon, transfer onion to a plate and in the same fat sauté beef until it changes colour. Return onion to the pan and mix well.

When split peas are soft and almost mushy, drain them thoroughly and mix them with the beef and onion. Sprinkle in the cinnamon and season to taste with salt and freshly ground black pepper. Leave to cool.

Meanwhile, prepare the apples. Wipe them clean with a damp cloth and core them from the stem end to within about 1 centimetre (½ in) of the bottom. Core out some of the pulp as well, leaving thick shells firm and unbroken. Chop up the pulp.

Lightly butter a baking dish that will take all the apples in a single layer. Scatter the apple pulp evenly over the dish. Sprinkle with a pinch of cinnamon, if liked, and moisten with a few tablespoons of water.

Stuff the apples with a meat mixture and arrange them side by side on the chopped apple pulp. Dot each apple with a flake of butter. Bake in a moderate oven (350°F, 180°C, Mark 4) for 30 minutes.

Prepare a sweet-sour basting sauce for the apples by combining the vinegar and sugar with 150 millilitres (¼ pint) water. Bring to boiling point, stirring until sugar has melted, and simmer for 1 minute.

Spoon a little boiling sauce into the cavity of each apple. (Any sauce left over may be poured over the chopped apples lining the baking dish.) Continue to bake apples for about 15 minutes longer until they are quite soft. Crush the apple pulp and pan juices together, and serve as a sauce for the apples. Serve hot.

Serves 4—5

WALDORF SALAD

4 red-skinned dessert apples
3–4 tablespoons lemon juice
4–6 stalks celery
50 grams (2 oz) walnuts
100–125 grams (4 oz) raisins
150 millilitres (¼ pint) thick mayonnaise

Quarter, core and cut the apples into small dice without peeling them. Toss them with lemon juice to prevent discoloration. Halve celery stalks lengthwise if they are thick and slice them thinly. Coarsely chop the walnuts and rinse the raisins, draining them thoroughly.
In a large mixing bowl, combine the apples and their juices with the remaining ingredients. Add mayonnaise and toss lightly until well mixed. Pile into a bowl and serve.

Serves 4–6

BURGHUL PILAF WITH APPLES AND ALMONDS

1 large Spanish onion, finely chopped
50 grams (2 oz) butter
150 grams (5 oz) burghul wheat
3 tart, crisp eating apples, peeled, cored and diced
¼ teaspoon crumbled dried thyme
¼ teaspoon celery seed
1 tablespoon finely chopped parsley
400 millilitres (¾ pint) well-flavoured chicken stock
Salt and freshly ground black pepper

Almond garnish
6–8 tablespoons slivered blanched almonds
50 grams (2 oz) butter

In a large, heavy saucepan or flameproof casserole with a tight-fitting lid, sauté onion in butter until soft and golden. Stir in dry burghul and sauté for a few minutes longer over moderate heat until lightly coloured and aromatic. Add diced apples and continue frying, stirring, until each cube is coated with butter. Sprinkle with dried thyme, celery seed and parsley, and stir in chicken stock.

Bring to boiling point, cover pan tightly and either cook over the lowest possible heat for about 25 minutes until all the stock is absorbed or bake in a moderate oven (350°F, 180°C, Mark 4) for roughly the same length of time. When burghul is cooked, fluff it up with a fork and correct seasoning with a little salt or pepper if necessary.

Have almonds already sautéed until golden in butter. Serve burghul in a heated bowl with hot almonds and butter poured over the top.

Serves 4–6

SALADE PARISIENNE

2–3 large, crisp apples
2 large potatoes, boiled in their jackets
3 bulbs chicory
2 bananas
1 large stalk celery
3 carrots
50 grams (2 oz) nuts, coarsely chopped
50 grams (2 oz) raisins

Dressing
3 tablespoons salad oil
1 tablespoon lemon juice
Pinch of paprika
Salt and freshly ground black pepper

Make the dressing for the salad first. Put the oil, lemon juice, paprika, salt and black pepper to taste in a screw-top jar. Close tightly and shake well to emulsify dressing. Peel, core and dice the apples, and put them in a large bowl. Peel and dice the potatoes, and add them to the apples. Trim the chicory bulbs. Slice them thinly, together with the bananas and celery, and add them to the bowl. Shred the carrots coarsely and add them. Finally, add the nuts and raisins.

Toss all the ingredients gently with a large kitchen fork until well mixed. Shake the dressing up once more and mix it lightly but thoroughly into the salad. Correct seasoning and chill lightly until ready to serve.

Serves 6

Note: If you find the salad is rather 'dry', make up some more dressing in the same proportions and add it to the bowl (the exact amount needed will depend on the juiciness of the fruit and vegetables).

ITALIAN APPLE PANCAKE

2 tablespoons plain flour
Pinch of salt
150 millilitres (¼ pint) milk
1 tablespoon caster sugar
½ teaspoon vanilla essence
2 tablespoons raisins
2 eggs
2 small dessert apples
Butter, for frying
Sifted icing sugar or caster sugar, to serve

First prepare a thin batter by blending the flour and salt smoothly with the milk. Stir in sugar and vanilla essence.
Rinse raisins and put aside to drain in a sieve.
Beat eggs lightly in a large bowl and gradually beat in prepared batter. When smoothly blended, stir in drained raisins. Put aside while you quarter, peel, core and slice apples paper-thin. Stir apples into the prepared batter.
Bring out a thick frying pan 17.5–20 centimetres (7–8 in.) in diameter and a flat plate that is slightly larger. Grease plate lightly all over with butter. In the frying pan, melt a large knob (about 25 grams or 1 oz) butter and when sizzling, tilt pan around so that entire surface is coated.
Pour in batter, making sure apple slices and raisins are evenly distributed in the pan. Cook over low heat until pancake batter has set on top and pancake is a rich golden brown underneath.

Carefully turn pancake out upside down on to the buttered plate. Melt a small knob of fresh butter in the pan and slide the pancake back in to cook and brown the other side over low heat.

Invert pancake on to a heated serving dish.

Dredge with icing sugar or sprinkle with caster sugar and serve immediately, cut in wedges.

Serves 2 as a snack, 4 as a dessert course

SOUTHERN APPLE PIE

About 450 grams (1 lb) shortcrust pastry for a 2-crust, 22.5
centimetre (9 in.) pie
700–900 grams (1½–2 lbs) tart dessert or cooking apples
Lemon juice (optional)
125–175 grams (4–6 oz) white or soft brown sugar, or half and half
¼ teaspoon ground cinnamon
⅛ teaspoon grated nutmeg
2 tablespoons plain flour
Pinch of salt
50 grams (2 oz) seedless raisins
15–25 grams (½–1 oz) butter
Caster sugar

Start by lining a pie dish with shortcrust pastry,
overhanging the sides a little, and cutting out a lid to go
on top. Put aside until needed.

Quarter, peel and core apples, and slice them thinly into a
bowl. If they are not too sour, or are rather old and
tasteless, toss them with a tablespoon of lemon juice.

Blend sugar with spices, flour and salt, and sprinkle over apples. Add raisins and toss lightly until thoroughly mixed.

Pack apple mixture into pastry-lined pie dish. Dot surface all over with flakes of butter.

Lay pastry lid in position. Press all around sides to seal them, then flute rim attractively between forefinger and thumb. Cut slits in top crust.

Bake pie in a hot oven (450°F, 230°C, Mark 8) for 15 minutes, then reduce heat to moderate (350°F, 180°C, Mark 4), and continue to bake for 35–40 minutes longer, or until top crust is a rich golden colour and apples feel soft when prodded with a skewer through one of the slits.

Serves 6–8

SHERRIED APPLES WITH SPICED NUTS

6 large baking apples
Butter
150 millilitres (¼ pint) medium sherry
Whipped cream, to serve

Nut stuffing
50 grams (2 oz) butter
85 grams (3½ oz) soft brown sugar
1 teaspoon ground cinnamon
1 tablespoon medium sherry
6–8 tablespoons chopped walnuts

Prepare apples in usual way, coring them and peeling away about a third of the skin, starting from the top.

Prepare nut stuffing by creaming butter and brown sugar together until light. Beat in cinnamon and sherry until thoroughly blended, then stir in chopped nuts.

Fill apples with nut stuffing. Arrange them side by side in a buttered baking dish. Pour sherry evenly over the top.

Bake apples in a moderate oven (375°F, 190°C, Mark 5), basting frequently with pan juices, for 30–40 minutes, or until they feel soft when pierced lightly with a skewer.

Serve lukewarm or cold, with a bowl of whipped cream.

Serves 6

HOT APPLE CHARLOTTE

Butter and breadcrumbs for baking dish
12 slices bread
400 millilitres (¾ pint) milk
2 eggs
1 tablespoon sugar
8 apples, peeled, cored and sliced
2 tablespoons chopped candied peel
2 tablespoons chopped raisins
3 tablespoons icing sugar, sifted
1 teaspoon vanilla essence

Butter a deep baking dish (a soufflé dish is ideal) and coat it with breadcrumbs. Toast the bread slices lightly on both sides until golden but not browned.

Dip each slice of toast on both sides in a mixture of milk beaten with eggs and sugar, and put them aside on a plate. Mix the apples with the candied peel, raisins, icing sugar and vanilla.

Arrange four layers of toast and three layers of apples in the baking dish (ie, starting and ending with a layer of toast). Pour in any remaining milk mixture.

Bake the charlotte in a moderately hot oven (400°F, 200°C, Mark 6) for 45 minutes, or until the top is crisp and golden. Serve hot.

Serves 4—6

BAKED RICE AND APPLE PUDDING

---

225 grams (8 oz) short-grain rice
25 grams (1 oz) butter
Salt
50 grams (2 oz) raisins
1 – 2 tablespoons chopped crystallized orange peel
900 grams (2 lb) cooking apples or crisp, tart dessert apples
White or soft brown sugar
1 teaspoon vanilla essence
25 grams (1 oz) butter

Measure the rice in a mug (there should be a mugful) and put it in a pan. Add 3 times the volume (ie about 3 mugfuls) of boiling water, the butter and a generous pinch of salt; stir well and cook very gently with the lid half on the pan until the water has been absorbed and the rice is very soft. Stir in the raisins and peel, cover tightly and put aside while you prepare apples.

Peel, core and slice the apples very thinly into a bowl. Add sugar to taste, sprinkle with vanilla and toss well with a large fork.

Grease a large baking dish generously with some of the butter, and spread base with a third of the rice mixture. Cover with half of the apples, followed by half of the remaining rice, all the remaining apples (and their juices, if any) and the rest of the rice, spreading it out as evenly as possible with the back of your spoon.

Dot the top of the pudding all over with flakes of remaining butter and bake in a moderate oven (350°F, 180°C, Mark 4) for 45–50 minutes, or until surface is crusty and golden brown. Serve hot.

Serves 6

RUSSIAN SOURED APPLES

4.5–5 kilograms (10–11 lbs) apples
About 25 grams (1 oz) fresh blackcurrant leaves
100–125 grams (4 oz) rye flour
100–125 grams (4 oz) sugar or liquid honey
1 teaspoon salt

Wash and drain apples thoroughly. Scrub your pot or urn and line base with a layer of fresh blackcurrant leaves. On this place a layer of apples, fitting them tightly side by side. Cover with more blackcurrant leaves. Continue in this manner until apples have all been packed into pot, ending with a layer of leaves.

Prepare preserving liquor. In a large bowl, work rye flour to a smooth paste with a little cold water. Stirring rapidly, pour in 4.4–5 litres (8 or 9 pints) of boiling water, making sure flour does not go lumpy. Stir in sugar or honey and salt, and put aside until cold.

Pour cold liquor over apples to cover them completely. Then place wooden board on top and weight it down with a muslin-wrapped stone to prevent apples floating to the top. They must at all times remain submerged. Allow apples to marinate in a cool, dry place (such as a garage, outhouse or cellar) for at least 4 weeks before tasting them. They will keep for several months, at least until the following spring. Should mould form on the liquor and turn the stone slimy, skim the liquor, scrub the stone and the muslin separately, and scald them with boiling water.

SWISS APPLE TART

One 22.5 centimetre (9 in.) sweet shortcrust pastry case, pre-baked
5–6 firm, tart, dessert apples
50 grams (2 oz) butter
1 teaspoon vanilla essence
3 small eggs
100–125 grams (4 oz) caster sugar
300 millilitres (½ pt) single cream
Sifted icing sugar, to serve

Peel, core and slice apples thickly. In a large, heavy frying pan, simmer apple slices in foaming butter until they are golden and just tender but not mushy, turning them over gently once or twice with a large spatula. Sprinkle with vanilla and, using the spatula, carefully transfer the slices to the pre-baked pastry case in a neat, even layer.

Beat eggs with sugar until thick and lemon-coloured. Add cream and mix well. Place tart tin back on baking sheet. Cover apples with egg mixture, pouring it in quite slowly over the back of a wooden spoon to avoid disturbing apples.

Return tart to the oven for 30 minutes, or until custard has set, no longer trembles when baking tin is pushed, and is lightly coloured on top. Cool to lukewarm and dust with sifted icing sugar before serving.

Serves 6–8

APPLE JELLIES

2.7 kilograms (6 lbs) apples
1 lemon
450 grams (1 lb) sugar per 550 millilitres (1 pint) of juice

Chop up apples but do not peel or core them. Finely grate
rind of the lemon, sprinkle it over the apples in a
preserving pan and cover with 2.7 litres (5 pints) water.
Bring to the boil and cook until apples are very soft and
pulpy. Tip the contents of the pan into a jelly bag and
allow to drip dry, or leave the apple pulp to drain in a hair
sieve.
Measure juice and weigh out 450 grams (1 lb) sugar per
550 millilitres (1 pint).
In the rinsed-out pan, combine the apple juice with the
sugar and strain in the juice of the lemon. Bring to the boil
slowly, stirring frequently, until sugar has dissolved. Then
boil rapidly until setting point is reached, skimming scum
from the surface.
Cool slightly, pot and cover.

FRESH APPLE RELISH

450 grams (1 lb) dessert apples
1½ – 2 tablespoons caster sugar
1 tablespoon crumbled dried mint
Cider or white wine vinegar

Peel the apples and grate them coarsely into a bowl containing the sugar, mint and about 3 tablespoons vinegar.

Toss lightly but thoroughly until well mixed (this will prevent the apples discolouring).

Taste and add more sugar or mint if liked, and if necessary pour in a little more vinegar so that apples are thoroughly moistened.

APPLE CHUTNEY

1.8 kilograms (4 lbs) peeled, cored and quartered apples
450 grams (1 lb) seedless raisins, chopped
450 grams (1 lb) dried dates, pitted and chopped
900 grams (2 lbs) brown sugar
2.2 litres (4 pints) brown malt vinegar
50 grams (2 oz) mustard seed
50 grams (2 oz) ground ginger
1 tablespoon salt
7 grams (¼ oz) cayenne pepper, or to taste
15 grams (½ oz) garlic cloves, peeled and very finely chopped

Combine all the ingredients in a preserving pan and mix well. Stir constantly over low heat until sugar has dissolved. Then bring to boiling point and simmer for between 1½ and 2 hours, stirring at an increasing rate as chutney thickens to prevent it scorching on the bottom of the pan.

Pour into a bowl and leave until the following day before potting and covering tightly.

Fills about 7 jars

APPLE CIDER SAUCE

25 grams (1 oz) butter
3 tablespoons flour
550 millilitres (1 pint) apple cider, warmed
¼ teaspoon ground cloves
¼ teaspoon ground ginger
Honey or brown sugar

Melt the butter in a saucepan over gentle heat. Mix in the flour and cook, stirring, for 1–2 minutes. Slowly stir in the apple cider and add cloves and ginger. Bring to the boil, still stirring, then remove from heat. Sweeten with honey or brown sugar.

APPLE AND GREEN TOMATO CHUTNEY

450 grams (1 lb) green tomatoes
450 grams (1 lb) apples, quartered and cored
450 grams (1 lb) onions
225 grams (8 oz) sultanas
225 grams (8 oz) brown sugar
25 grams (1 oz) salt
15 grams (½ oz) ground ginger
15 grams (½ oz) cayenne pepper
½ teaspoon ground cloves
550 millilitres (1 pt) brown malt vinegar

Put tomatoes, apples, onions and sultanas through a mincer, together with brown sugar, and put in a preserving pan. Blend salt and spices smoothly with some of the vinegar. Mix with remaining vinegar, pour into preserving pan and mix well.

Place pan over low heat and stir until sugar has melted and mixture comes to boil. Then boil gently for about 1 hour, stirring more and more frequently as chutney thickens to prevent it catching on the bottom of the pan. Cool.

Pot and cover.

Fills about 4 jars

OLD-FASHIONED APPLE BUTTER

2.7 kilograms (6 lbs) apples
1.1 litre (2 pints) sweet cider
Sugar
1½ teaspoons ground cinnamon
½ teaspoon ground cloves

Rinse apples and chop them up. There is no need to peel or core them.

Put them in a preserving pan with cider and barely enough water to cover. Bring to boiling point and simmer the apples until soft and pulpy, stirring occasionally.

Rub contents of pan through a sieve, discarding peels, pips and other debris.

Weigh pulp and return it to rinsed-out pan. Weigh out 350 grams (12 oz) sugar for every 450 grams (1 lb) of apple purée.

Bring purée to simmering point, stirring frequently. Add prepared sugar and spices. Stir until sugar has dissolved, then simmer, stirring regularly at an ever-increasing rate as purée thickens to ensure that it does not catch on the bottom of the pan. The butter is ready when no surplus liquid can be seen collecting around the edges of a small mound of purée which you have spooned out on to a saucer.

Pot and cover.

BREAKFAST
APPLE SHAKE

2 tart apples, peeled and cored
300 millilitres (½ pint) milk
Pinch ground cinnamon
1 tablespoon honey
Ice cubes (optional)

Chop the apples and put in the bowl of an electric blender
with the milk, cinnamon and honey.

Serves 2–3

Note: Skimmed milk can be substituted for full fat milk to increase the
nutritional value of this drink.

APPLE AND ALMOND PACK

This is good for rough skin (on elbows and knees, for instance), and can also be used as a face pack for dry skin. Peel and grate a firm apple. Add about the same amount of ground almonds, and mix in milk (or cream) until you have a slightly gritty but spreadable paste. Spread the paste on your elbows, knees (or anywhere else with less than smooth skin) or face, and leave to dry. Rub off from arms and legs – preferably over a bath – or wash off from your face with warm water.

APPLE-BASED REMEDIES

A diet of grated apple, in moderation, will often clear up diarrhoea in a couple of days. Use slightly unripe fruit, and peel them first.

Cider vinegar has been hailed as a cure for almost every ill under the sun. Most people are justifiably sceptical about this claim, but it is an excellent blood cleanser and helps to eliminate a vast variety of toxins. Up to two teaspoons in a tumbler of water, two or three times a day, will help digestion and promote energy.

MEASUREMENTS

Quantities have been given in both metric and imperial measures in this book. However, many foodstuffs are now available only in metric quantities; the list below gives metric measures for weight and liquid capacity, and their imperial equivalents used in this book.

WEIGHT

25 grams	1 oz
50 grams	2 oz
75 grams	3 oz
100 – 125 grams	4 oz
150 grams	5 oz
175 grams	6 oz
200 grams	7 oz
225 grams	8 oz
250 grams	9 oz
275 grams	10 oz
300 grams	11 oz
350 grams	12 oz
375 grams	13 oz

400 grams	14 oz
425 grams	15 oz
450 grams	1 lb
500 grams (½ kilogram)	1 lb 1½ oz
1 kilogram	2 lb 3 oz
1.5 kilograms	3 lb 5 oz
2 kilograms	4 lb 6 oz
2.5 kilograms	5 lb 8 oz
3 kilograms	6 lb 10 oz
3.5 kilograms	7 lb 11 oz
4 kilograms	8 lb 13 oz
4.5 kilograms	9 lb 14 oz
5 kilograms	11 lb

LIQUID CAPACITY

150 millilitres	¼ pint
300 millilitres	½ pint
425 millilitres	¾ pint
550 – 600 millilitres	1 pint
900 millilitres	1½ pints
1000 millilitres (1 litre)	1¾ pints
1.2 litres	2 pints
1.3 litres	2¼ pints
1.4 litres	2½ pints
1.5 litres	2¾ pints
1.9 litres	3¼ pints
2 litres	3½ pints
2.5 litres	4½ pints

OVEN TEMPERATURES

Very low	130°C, 250°F, Mark ½
Low	140°C, 275°F, Mark 1
Very slow	150°C, 300°F, Mark 2
Slow	170°C, 325°F, Mark 3
Moderate	180°C, 350°F, Mark 4
	190°C, 375°F, Mark 5
Moderately hot	200°C, 400°F, Mark 6
Fairly hot	220°C, 425°F, Mark 7
Hot	230°C, 450°F, Mark 8

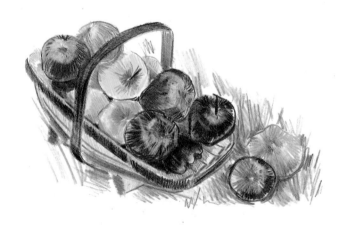